# YOUR KNOWLEDGE HAS VALUE

- We will publish your bachelor's and master's thesis, essays and papers

- Your own eBook and book - sold worldwide in all relevant shops

- Earn money with each sale

Upload your text at www.GRIN.com and publish for free

**Bibliographic information published by the German National Library:**

The German National Library lists this publication in the National Bibliography; detailed bibliographic data are available on the Internet at http://dnb.dnb.de .



**Imprint:**

Copyright © 2012 GRIN Verlag, Open Publishing GmbH
Print and binding: Books on Demand GmbH, Norderstedt Germany
ISBN: 9783668431553

**This book at GRIN:**

http://www.grin.com/en/e-book/358659/piggery-farmers-awareness-of-the-implications-of-the-use-of-aflatoxins

**Samuel Ekwu, A. E. Onyimonyi, S. O. Ugwu**

# Piggery Farmers' Awareness of the Implications of the Use of Aflatoxins Contamination of Feedstuffs in the Humid Tropics

**An Evaluation**

GRIN Publishing

# An Evaluation of Piggery Farmers' Awareness of the Implications of the Use of Aflatoxins Contamination of Feedstuffs In the Humid Tropics

Onyimonyi, A. E, Ugwu, S.O. C and Ekwu, U. S.
Department of Animal Science, University of Nigeria, Nsukka, Nigeria.

**Abstract**

*This study was conducted to evaluate piggery farmers'awareness of aflatoxins contamination of feedstuffs and its implications in pig production in Nsukka zone. A survey research was conducted using a structured questionnaire which was validated by an agricultural measurement and evaluation experts. The questionnaire was used to collect data from the piggery farmers. The data collected were analyzed using descriptive statistics option ofGenstat computer package (Discovery edition 3). The result of this study revealed that piggery farmers are aware of aflatoxins infestation, butknow little about its economic and health implications on pig production and pork consumers. They also indicated that they are adopting various local and technical methods such as avoiding damp environment, sprinkling ashes and saw-dusts on top of the feedstuffs, in order to reduce the incidence of aflatoxins development in the feedstuffs they use to feed their pig stocks. They stated that they were not aware that feeding pigs with aflatoxins contaminated food reminants and feeds poses a serious health risk to both the pigs and pork consumers, as they are interested in minimizing the cost of production rather than the welfare of the pigs and pork consumers. Pigs fed with aflatoxin contaminated feedstuffs, have stunted growth, poor feed intake and liver damage, when consumed by humans cause diseasessuch as liver cancer, kidney and nervous disorders, cirrhosis and mutagenswhich have been proven by researchers. Hence the need to ascertain the extent of piggery farmers' awareness of toxins, which will serve as an indicator on the level of risks exposure of both pigs and pork consumers in the zone. It was therefore recommended that more awareness should be created to both the crop and livestock farmers on the economic and health implications of aflatoxins in the food chain. Farmers should be strongly encouraged to adopt a good preventive and control strategies both in the field, storage and handling of feeds and feed ingredients in order to avert a possible outbreak of aflatoxicosis if uncheck-mated.*

**Keywords:**

Aflatoxins, aflatoxin contamination, feedstuffs, pig, piggery farmers, humid tropics

**Introduction**

Babatunde and Hamzat (2005) these alternative feedstuffs have proved valuable in supporting the performance of livestocks and poultry. According to Ugwu, *et al.,* (2008) the main factors militating against the rapid expansion of animal production industry in Nigeria is the problem of inadequate supplies of feedstuffs at economic prices. The scarcity and  high cost of conventional feedstuffs is largely responsible for the present high price of  finished animal products such as eggs, meat and milk (Adesehinwa*et al.,* 2011; Rhule,1999). Ijaiya*et al.,* (2004) pointed out that feed cost is perhaps the most expensive input in intensively reared stocks and constitutes about 70 - 80% of the real cost of  animal production.Adesehinwa (2008) noted that pigs convert a variety of feeds and agro-industrial by-products into meat for human consumption.Serres (1992) stated that pigs are  highly prolific and very efficient in converting feed nutrients into high quality animal protein. Awan (2001) reported the ability of moulds to produce toxins that has deleterious to animals and human health are common and widespread. Odoemelam and Osu, (2008) reported that toxigenic moulds have been found during growth, harvest and storage of diferent foods and feeds and other agricultural produce. Farombi (2006) stated thattoxigenic fungal contaminate a large number of dietary staples and agricultural produce such as rice, corn, cassava, peanuts and spices, while noting that humans and animals are exposed to aflatoxins by consuming contaminated foods and feeds. Abarca*et al.,*(2001) stated that there are five major agriculturally important fungal toxins of economic importance namely; aflatoxins, deoxynivalenol, ochratoxin, zearalenone and fumunism, which are produced by fungi invasion of agricultural produces and feed ingredients under favourable conditions. According to Sashidhare*et al.,*(1992), the high incidence of mycotoxins contamination could be due to physical status of grains, the moisture content, temperature, oxygen and the amount of carbon (iv) oxide in the atmosphere. These factors are reported to influence the rate of infestation and proliferation of mouldsin agricultural produce and grains especially under storage condition.Aflatoxicosis in pigs reduces feed intake, reduce weight gain, reduce growth rate, immunosuppressive including antibody and interleukin production, causes liver damage and hemorrhage (Sun and Chen, 2003; Carvajah*et al.,* 2003; Lu, 2003 ;Creepy, 2002). Porks consumers'are exposed to health risks such as cancer, kidney and nervous disorders, cirrhosis and mutagens (Kuilman*et al.,*1998). These toxins are wildly implicated in causing health risks to human due to its carcinogenic, mutagenic and teratogenic properties in persons consuming pork products that are high in aflatoxins concentration over time (Sun and Chen, 2003; Kuilman*et al.,* 1998).The purpose of this study was to evaluate the extent of piggery farmers' awareness of the economic and health risk of feeding aflatoxinscontaminated feeds and feed ingredients in pig production.

## Material and Methods

### Location and duration of the study

This study was carried out in the six (6) local government areas that make up Nsukka zone namely: Nsukka, Igboeze North, Igboeze South, Igbo-etiti, Udenu and Uzo-uwani LGAs of Enugu State, South Eastern Nigeria. The climate of the study area is typically tropical with relative humidity ranging from $65 - 85$ %. The average diurnal minimum temperature ranges from $22^{\circ}C - 24.7^{\circ}C$ while the average maximum temperature ranges from $33^{\circ}C - 37^{\circ}C$ (Energy Centre, UNN, 2008) with annual rainfall ranging from $1680 - 1700mm$. The rainy season is between April – October and the dry season is between Novembers – March. This study was carried out in two major piggery farms in each of the six (6) sampled local government areas in Nsukka zone of Enugu State. The study lasted for thirty two weeks from December, 2011 - July, 2012.

### Population of the study

The population of the study comprised of all piggery farmers in Nsukka zone of Enugu State. Three major piggery farmers were purposively sampled in each of the six local government areas that make up Nsukka zone in this study, with a total number of eighteen (18) piggery farmers sampled in this study.

### Experimental Procedure/Sampling

Three major piggery farmers from each of the six (6) local governments areas in Nsukka zone were administered a structured questionnaire. These questionnaires where administered by the researcher, in cases where a given question(s) is not understood by the piggery farmer, the researcher explained in the native dialect and also fill in the responses if the piggery farmer is unable to write. The administered questionnaire was retrieved immediately by the researcher.A pilot aflatoxins analysis of the pig feedstuff samples was carried out using AOAC (2005) method of thin-layer chromatography (TLC) with some modifications according to Opadukun (1979) method, used by Odoemelam and Osu, (2008) and modified according to Ono *et al.,* (2010). This pre-test is to ascertain the presence of aflatoxins in the mould infested feedstuff samples collected being used by piggeryfarmers in feeding pigs.

## Data Collection

The data were collected using a structured questionnaire. The questionnaire was validated by two experts in instrument measurement and evaluation from the department of Agricultural Economics, University of Nigeria, Nsukka.

## Statistical Analysis

The collected data from questionnaires were analyzed, using descriptive statistics such as frequency and percentages.Genstat computer package (Discovery edition 3) software was used to analyze the data.

## Results and Discussion

The results of the study are presented below; table 1 shows thevarious responses of piggery farmers

**Table 1: The results of the questionnaire responses of piggery farmers**

| Questionnaire Responses of Piggery Farmers | Frequency (F) | Percentage (%) |
|---|---|---|
| **1. Marital status** | | |
| Married | 11 | 95 |
| Single | 1 | 5 |
| Divorced | - | - |
| Total | 12 | 100 |

**Educational Qualification**

| | | |
|---|---|---|
| WASC/O 'Level | 11 | 95 |
| OND | - | - |
| HND | - | - |
| Degree | 1 | 5 |
| Total | 12 | 100 |

**1.  Gender**

| | | |
|---|---|---|
| Male | 12 | 100 |
| Female | - | - |
| Total | 12 | 100 |

**4. Number of Pigs reared in the farm**

| | | |
|---|---|---|
| < 20 | 6 | 50 |
| 20 – 50 | 5 | 42 |
| > 50 | 1 | 8 |
| Total | 12 | 100 |

**5. The feedstuffs used  by piggery farmers to feed pigs in the farm**

- Bambara nut waste
- Brewer spent grains
- Cassava leaves
- Cassava peels
- Cassava roots
- Guinea and elephant grasses and other legumes shrubs.
- Hays
- Left over foods from homes, eateries and restaurants
- Mash water, mash palm kernel water
- Palm kernel cake
- Pawpaw fruits
- Pawpaw leaves
- Plantain leaves
- Plantain peels
- Yam peels

**6. Main Sources  of feedstuffs from where the pig farmers procure feeds**

- Family Farm
- Garri Mills/Plants
- Palm Oil Mills
- Livestock Feed Dealers
- Eateries and Cafeteria
- Bambara nut milling plants
- Nigeria Breweries

**7. Frequency of  Sourcing of  these feedstuffs**

| | | |
|---|---|---|
| | 4 | 34 |

| | | | |
|---|---|---|---|
| Weekly | | 6 | 50 |
| Two weekly | | 1 | 8 |
| Monthly | | 1 | 8 |
| Quarterly | | 12 | 100 |
| Total | | | |

## 8. The quantity of the feedstuffs purchased in kg/tons

| | | | |
|---|---|---|---|
| Cassava peels | **10 – 25** | 6 | 50 |
| | **26 – 41** | 3 | 25 |
| | **42 – 57** | 3 | 25 |
| | | | |
| Bambara nut waste | **10 – 25** | 2 | 16.7 |
| | **26 – 41** | 2 | 16.7 |
| | **42 – 57** | 8 | 66.7 |
| | | | |
| Palm kernel cake | **10 – 25** | 3 | 25 |
| | **26 – 41** | 2 | 16.7 |
| | **42 – 57** | 7 | 58.3 |
| | | | |
| Brewer spent grains | **10 – 25** | - | - |
| | **26 – 41** | - | - |
| | **42 – 57** | 12 | 100 |

## 9. How these feedstuffs are stored in the pig farms

- Sack bags in roofed stores
- Cemented dug pit with long over hang roof
- Heap covered with sack bags with heavy log of wood
- Drums with lid

## 10. Susceptibility of Pig breeds to mould contaminated feedstuffs

| | | |
|---|---|---|
| Yes | 11 | 95 |
| No | 1 | 5 |
| Total | 12 | 100 |

## 11. Ingredients added to the feedstuffs to improve its nutrients availability

| | | |
|---|---|---|
| Salt | | |
| Premix | 7 | 58 |
| Both | 3 | 25 |
| Total | 2 | 17 |
| | 12 | 100 |

## 12. Mould incidence/infestation of feedstuffs fed to pigs

| | | |
|---|---|---|
| Yes | | |
| No | 11 | 95 |
| Total | 1 | 5 |
| | 12 | 100 |

## 13. Feedstuffs that develops moulds faster than others

| | | |
|---|---|---|
| Cassava peels | | |
| Bambara nut waste | 1 | 8 |
| Palm kernel waste | 5 | 42 |
| Brewery spent grain | 3 | 25 |
| Total | 3 | 25 |
| | 12 | 100 |

**14. Signs/Symptoms shown by the pigs when fed mould contaminated feeds**

- Fever
- Feed refusal
- Weakness
- Vomiting
- Death
- 

**15. How these symptoms affects the pigs performance**

| | | |
|---|---|---|
| • Weight loss | 1 | 5 |

**16. Preventive measures adopted to reduce mould infestation of the feedstuffs**

- Preventing water from wetting the stored feedstuffs in the bags.
- Spreading of wood ash on top of the feedstuffs, while in sack bags
- Covering the brewer spent grain with rice husk
- Salt is also added as a preservative substance in the stored feedstuffs
- 

**17. The season/period of the year mould infestation of feedstuffs are prevalent**

| | | |
|---|---|---|
| Jan – Mar | | |
| Apr – Jun | - | - |
| Jul – Sept | 4 | 33 |
| Oct – Dec | 8 | 67 |
| Total | - | - |
| | 12 | 100 |

---

Source: field survey, 2012

Graphical Representation of the Results of the Responses of Piggery Farmers, on questions withMultiple Answers.

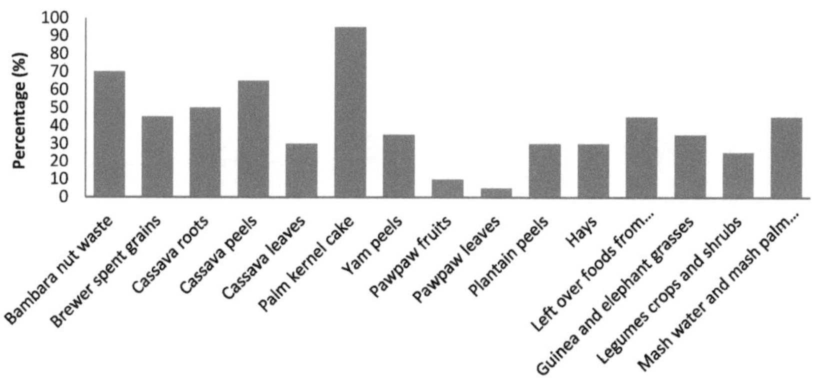



Fig 4: Bar chart showing the various feedstuffs used by piggery farmers in feeding pigs in their farms

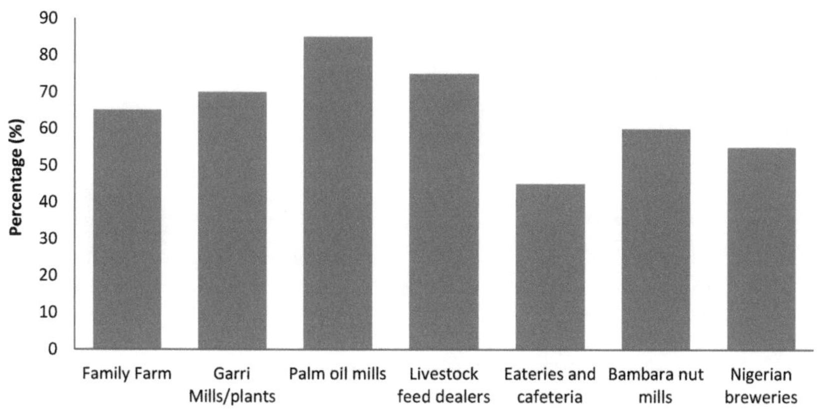



Fig 5: Bar chart showing the main sources of feedstuffs from where the pig farmers procure their feeds from.

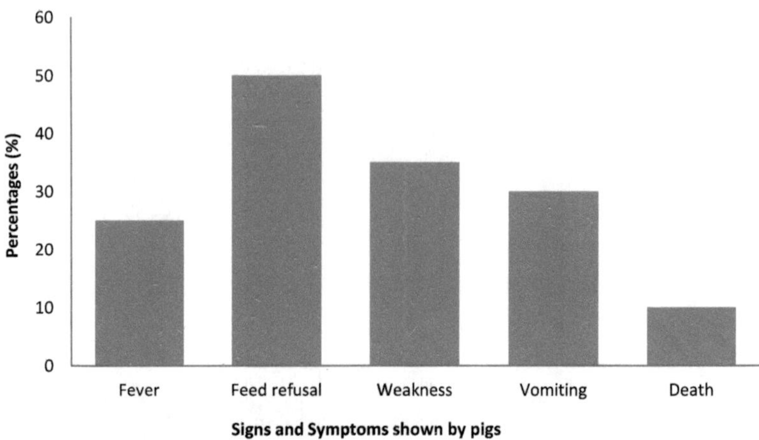

Fig 6: Bar chart showing the signs/ symptoms shown by pigs when they are fed mould
    contaminated  feeds

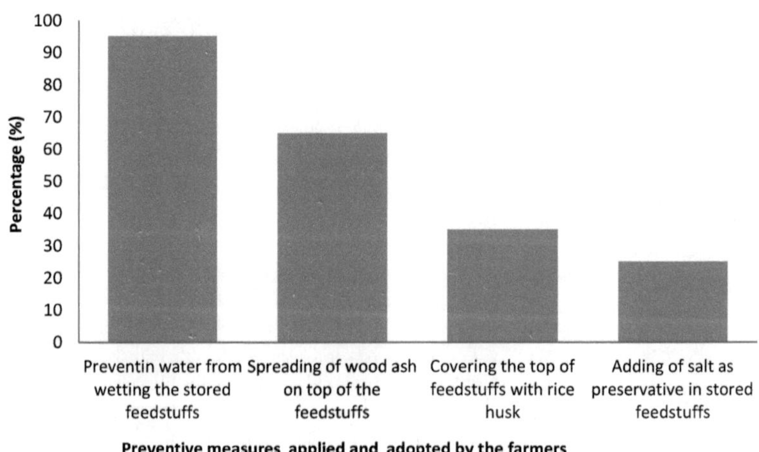

Fig 7: Bar chart showing the preventive measures adopted to reduce mould infestation of
    feedstuffs

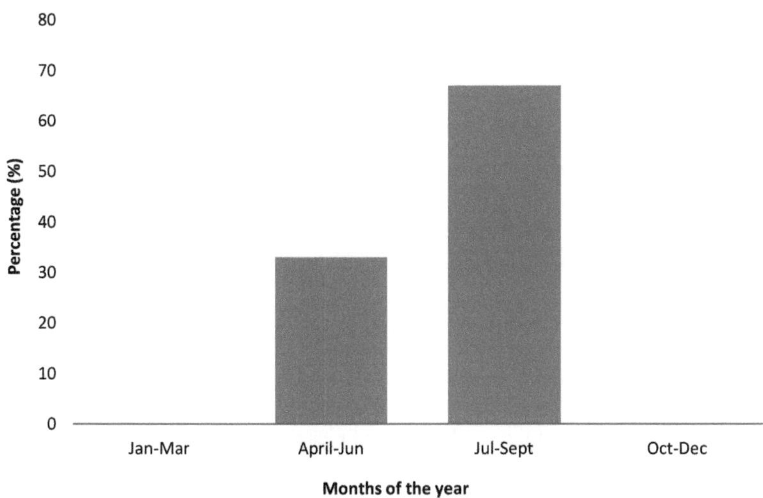


Fig 8: Bar chart showing the season/ period of the year mould infestation of feedstuffs are prevalent

*Source:* Field Survey, 2012.


**Discussion of the Questionaires Responses of Piggery Farmers**

***Marital status***: The results of the responses of piggery farmers on their marital status shows that most of the piggery farmers are married men i.e. 95% while one is yet to married which represents 5%. This result is in agreement with Ezidimma (2001), he noted that married people have the responsibility for the provision, processing and marketing of food items for the household. They added that piggery farming serves as a source of income to train their children in schools and meet other financial needs of the family. This indicated that piggery farming in this zone is not commercialized, and is perceived as an alternative source of income for their families and the family members in return give assistance in the small scale subsistence farming of both crops and animal farming.

***Educational qualifications***: The results of the responses indicated that majority which represents 95% of the piggery farmers possess WAEC i.e. O' Level and only one of them which represents 5%, had higher national diploma (HND). This implies that piggery enterprise in Nsukkazone  are dominated by O' level holders. This low level of education may affect their attitude and perceptions towards adopting new and modern methods of pig husbandry. This was in agreement with Uddin (2013) and Ajala (1992) they stated that majority of the farmers has low level of education, which has positive correlation with their acceptance and adoption of agricultural innovative techniques.

***Gender***: The results of the responses shows that all the piggery farmers sampled were all males which represents 100%. This impliesthat piggery farming in Nsukka zone is dominated by males. This finding is in agreement with Ajala (1992), he reported that majority of males are involved in livestock farming than females. This was supported by Achike (2002), who stated that more males were involved in farming than women in Enugu State.

This may also be as a result of the sturdy nature of pigs and their handling demands, which the females cannot easily cope with, rather prefer buying and selling at the major market days.This is in agreement with Adesiji*et al.,* (2012), they stated that males are more capable of doing tedious work, associated with farming than the females.

***Number of pigs reared in the farm***: The results of the responses shows that half i.e. 50% of the piggery farms sampled rear less than 20 matured pigs in their farms, while 42% have between 20 -50 pigs in their farms and only 8% have pig stock above 50 in number. This implies that piggery farming in this zone is not commercialized but rather are on subsistence basis and as a source of income and food for the household. By minimizing the number of pig stock they rear in their farms, it help them to minimize cost of feed, as more number of pigs implies more demand for feeds/feeding stuffs, medication and housing. This finding was in agreement with Oyekale (2009), he affirm that small scale farmers operate at subsistence level, and small size of cultivated land and small number of animal stocks being reared, is a characteristics feature of subsistence farming.

***Feedstuffs used by piggery farmers to feed pigs in their farms***: The results of the responses from piggery farmers were multiple, as there are many feedstuffs that they used to feed their pigs depending on theiravailability, quantity and cost. However palm kernel cake, bambara nut waste, cassava peels, cassava root, brewer spent grains are the major feedstuffs, frequently used in this areas to feed pigs, while others listed feedstuffs are moderately used to supplement these major feedstuffs in order to reduce cost or in times of scarcity. This was in support of Babatunde and Hamzat, (2005) and Adesehinwa, *et al.,*(2003)findings; they affirmed that industrial and agricultural by-products are good alternative feedstuffs for sustainable livestock production.

***Main source of feedstuffs from where the pig farmers procure feedstuffs:*** The results of the responses shows that the major sources of the various feedstuffs used to feed pigs in this zone are major markets, palm oil mills, garri mills, bambara nut mills, breweries industries and livestock feed dealers, are where most of the piggery farmers used to source their feedstuffs from. The piggery farmers always make their purchases from the actual producers of these feedstuffs and not from wholesale resellers who will add their own margin to make profit except in emergency situations. However it gives the piggery farmers the opportunity to book ahead of time from the mills, before their feedstuff stock finishes. This result was in support ofFatufe*et al.*(2007); Adesehinwa, *et al.,*(2003) and Onyimonyi (2002), they stated that the use of alternative feedstuffs support livestock production with appreciable reduction in the cost of production.

***Frequency of sourcing pigs' feedstuffs:*** The results of the responses shows that 50% of the piggery farmers which are the majority use to source their feedstuffs in two weeks interval, while 34% of the piggery farmers source their feedstuffs on weekly basis and 8% of the piggery farmers source their feedstuffs on monthly basis and quarterly respectively. This implies that the piggery farmers in the area mostly source for their feedstuffs between one and two weeks interval. This will make the feedstuffs not to stay too long in the farm before they are fed to pigs, as feedstuffs staying longer than this may give room for and encourage mould development and growth with the resultant aflatoxins contamination of the feeds. However poor handling and storage techniques/methods may exposure these feedstuffsto mould contamination and resultant aflatoxins production within this short time interval. This is in agreement with the findings of Whitlow (2011) and Ozsoy*et al.,* (2005) they reported that the incidence of aflatoxins concentration level may be dependent on bad storage of feed in farms and feed factories, adding that it could also depend on poor management, storage and warehousing in situation with humidity higher than 12% and temperature of $20^{o}$C.

***Quantity of feedstuffs purchased in kg/tons:*** The results of the responses shows that 50% of the piggery farmers  purchase between 10 - 25kg of cassava peels, while 25% purchase between 26 - 41kg and also 25% purchase between 42 - 57kg of cassava peels. Bambara nut waste 16.7% of the piggery farmers purchase between 10 - 25kg or 26 - 41kg while 66.7% of piggery farmers purchases 42 - 57kg. Palm kernel cake 25% of the piggery farmers purchase between 10 - 25kg, while 16.7% purchase between 26 - 41kg and 58.3% purchases between 42 - 57kg of palm kernel cake. Brewer spent grains 100% of the piggery farmers purchase 42-57kg and above. The usually purchase brewer spent grain from 9th mile with 911 full lorry load. This is in agreement with Oyekale (2009), he noted that livestock farming in most part of southern Nigeria are  small scale farms, operate at subsistence level, thereby the number of animals reared determines the quantity of feeds to be purchased.

***How these feedstuffs are stored in the piggery farms:*** The results of the responseson the various methods the piggery farmers used to preserve and store their feedstuffs against contamination of any kind including mould prevention,  shows that 90% which is majority stores their feedstuffs in sack bags kept in a roofed store that is rat proof or alternatively stored in drums with cover; also heaping of feedstuffs such as cassava roots and peels and covering them with sack bags with heavy logs of wood placed on top of the heaps, this is to give room for fermentation of the cassava roots and peels, the purpose of doing this is to reduce hydrogen cyanide toxicity (Onyimonyi, 2002; Adesehinwa, *et al.,* 2003; Du Thanh and Preston, 2005). Also cemented dug pit of 5 fts deep with two blocks above the ground level long over hand roofs are mostly used to store brewer spent grains.

***Susceptibility of pig breeds to mould contaminated feedstuffs:*** The results of the responses shows that 95% of the piggery farmers agree that pigs are vulnerable to mould contaminated feedstuffs, while 5% neither agree nor disagree with pig vulnerability to mould infested feedstuffs. This implies that  majority of the piggery farmers are aware of the vulnerability of pigs to mould contaminated feeds and the implications of feeding mould contaminated feeds to pigs.This is in accordance with Lu, (2003) and Sun and Chen, (2003) they stated that monogastrics animals are more susceptible to the effect of aflatoxins than ruminants.

***Ingredients added to the feedstuffs to improve its nutrient availability:*** The results of the responses shows that 58% of the piggery farmers use to add salt in pig's feeds for taste and palatability, 25% add premix while 17% add both salt and premix to the pig feedstuffs as a way to improve either the nutrient availability of the feedstuffs or for taste/palatability of the feeds to the pigs especially when such feeds are newly introduced to the pigs for the first time.

***Mould incidence/infestation of feedstuffs fed to pigs:*** The results of the responses shows that 95% of the sampled piggery farmers agreed that they feed mould contaminated feedstuffs to their pigs, while 5% neither agree nor disagree that they used to feed their pigs with mould infested feedstuffs. This implies that in spite of the piggery farmers being aware of the vulnerability/susceptibility of pigs to mould infested/contaminated feedstuffs, they still fed their pigs with such feeds, and this may be as a result of low level of income of the piggery farmers in order to minimize/reduce thecost of production, thereby sacrificing/endangering pigs' health and the health of pork consumers. This confirms the pilot investigation prior to this study, which revealed that piggery farmers feed their pigs with all sort of  toxic and contaminated materials. This was in accordance with Reddy and Farid, (2012), they stated that mould contaminates majority of agricultural crops especially in the field before they are harvested and are used as feeding stuffs by farmers.

***Feedstuffs that develop moulds faster than others:*** The results of the responses shows that 8% of the piggery farmers indicated cassava peels, 42%indicated bambara nut waste, 25% indicated palm kernel cake and also 25% indicated brewer spent grains. This implies that bambara nut waste were implicated by most of the piggery farmers to develop fungal/mouldcolonization faster than other feedstuffs while cassava peels are the least indicated to be capable of developingmould on storage, this may be as a result of the hydrogen cyanide content of the cassava peels which may interfere with mould growth and development.This was in agreement with Du Thanh and Preston(2005) and Onyimonyi, (2002).

***Signs and symptoms shown by pigs when fed mould contaminated feeds:*** The results of the responses shows that 50% of the piggery farmers indicated feed refusal, 35% indicated weakness, 30% indicated vomiting,25% indicated fever while 10% indicated death as the signs and symptoms exhibited by pigs when they are fed or exposed to mould contaminated feedstuffs. This is in agreement with Lin, *et al.,* (2004) ; CAST (2003); Carvajah, *et al.,* (2003) and Creppy (2002) they reported that pigs consuming aflatoxins contaminated feeds shows the above signs and symptoms over a given period of time depending on the quantity consumed over time and the length of exposure.

***How does these symptoms affect the pigs:*** The result of the responses shows that 5% of the piggery farmers indicated weight loss, while the rest 95% indicated no identifiable symptoms. Since the effects are not acute but gradual, hence the observable symptoms shown by the exposed pigs are not easily noticed by these farmers.This further indicated the economic importance of feeding pigs with mould contaminated feeds. This makes the exposed pigs to takes longer time in attending market weight; this is not economical to the piggery farmers and pig enterprise at large. This agrees with Ananth and Farid (2012), Dohlma (2004) and Sun and Chen, (2003) they reported that aflatoxins have huge economic implications which includes reduce feed intake, slower growth rates, loss of income, increased cost of production, animal death and other adverse health implications in livestocks, in this case pigs and pork consumers. This was in accordance with Akbar andMajid, (2012); Ozsoy*et al.*(2005)and CAST, (2003) reports that aflatoxins causes enormous economic implications.

***Preventive measures adopted to reduce mould infestation of the feedstuffs:*** The results of theresponses shows a multiple responses from the piggery farmers. However 90% majority stated that they prevent water from wetting the stored feedstuffs in the sack bags and use wooden slabs as a bedding on which the sack bags were kept, while 65% indicated that they use to spread wood ash on top of the feedstuffs in sack bags or in drums with cover, 30% indicated covering the feedstuffs especially brewer spent grains with rice husks and lastly 25% indicated the addition of salt as a preservative substance in the stored feedstuffs. All these preventive methods tends to achieve the following: prevent water from wetting or damping the store, thereby minimizing the moisture content, placing the sack bags on wooden slabs aids free air movement and circulation, spreading wood ash and rice husk on top of the feedstuffs, may help to reduce the moisture content of the feedstuffs since they are drier, hence will absorb water from the feedstuffs and also reduce the surface area exposure of the feedstuffs to microbial contamination.These are the locally devised preventive measures adopted by the farmers. This practice is in agreement with Dixon and Combellas, (1983), they reported that covering the surface of stored feeds with plastics and other non-contaminated materials will minimize surface spoilage and the length of time the feed material can be stored. This was in line with CAST (2003) and Gilbert and Vargas (2003), they stated that the conditions conducive for fungal development and growth are exposure to moisture, relative humidity of 80% or above, temperature range of  27 - 30$^{\circ}$C, ear injury caused by insects or birds as well as drought stress.

***The season/ period of the year mould infestation of feedstuffs are prevalent:*** The results of the responses shows that 33% of the piggery farmers indicated April-June, 67% indicated July-September while no one indicated neither January-March nor October-December. This implies that mould prevalent is more in the rainy season than in the dry season, the  agrees with the findings of Jewers*et al.,* (1986) and Pier, (1992), they reported seasonal variability in the aflatoxins concentration and that wet seasons favouraflatoxins (toxigenic fungal growth and development).  Also this results supports and is in agreement with the result of the laboratory analysis for aflatoxins in sampled feedstuffs this study which indicated that season have an effect on aflatoxins concentration  levels of feedstuffs with rainy season favouring fungal growth and development, which produces aflatoxins as its metabolites.


**Conclusion and Recommendations**

From the findings of this study, we concluded that  majorityof the piggery farmers are aware of the fungal contamination of their feedstuffs, however, most of them are ignorant of the economic importance and health risks it poses to the pigs and pork consumers. However they are adopting preventive measures in order to ameliorate the rate of contamination of their feeds and feed ingredients includingmould infestation/contamination. We therefore recommend that more awareness campaigns and strategies should be launch by Federal and state ministries of agriculture, research institutes, agricultural faculties in universities and agricultural extension service, food and drug regulators, feed industries to create awareness on fungal toxins especially aflatoxins, its economic implications, prevention and control practices, as well as its health risk to pork consumers both crop and livestock farmers. Farm visits, workshops and radio programmes are viable medium to disseminate this information to farmers.

**References**

Achike, A.I (2002).Gender Dynamics of Agricultural Technological Change Implication on Income Poverty Alleviation in Southeastern Nigeria. A research proposal submitted to the Africa Technology Policy Studies (ATPS) Network, Nairobi, Kenya, presented to the ATPS International Workshop held at Abuja, November 10 -16, 2002.

Adesehinwa, A. O.  K.,Aribido, S. O.,Oyediji, G. O and Obiniyi, A. A. (2003).Production Strategies for coping with the Demand and Supply of Pork  in some Peri-urban areas of Southwestern Nigeria. Livestock Research for Rural Development        15(10) http://cipav.org/irrd15/10/ades1510.htm

Adesehinwa, A. O. K. (2008). Energy and Protein Requirements of Pigs and the Utilization of Fibrous Feedstuffs in Nigeria: A Review. *Afr. J. of Biotech.*Vol. 7(25): 4798-4806.

Adesehnwa, A. O. K., Obi, O.O., Makanjuola, B.A., Oluwole, O. O and Adesina, M. A. (2011).Growing Pigs fed cassava peels based diet supplemented with or without F armazyme3000 proenx: Effect On Growth  on Growth, Carcass and Blood Parameters; *Afri. J. of Biotech.*Vol. 10 (14) pp.2791 - 2791.

Adesiji, G. B., Matanmi, B. M., Onikoyi, M. P and Saka, M. A. (2012). Farmers' perception of Climate change in Kwara State, Nigeria. World rural Observer 4 (2) 46 – 54. Retrieved from http://www.sciencepub.net/rural on 2nd August, 2013.

Ajala, A.A (1992). Factors Associated with Adoption of Improved Practices by Goat Producers in Southeastern Nigeria. Research Monograph no. 5, Department of Agricultural Extension, University of Nigeria, Nsukka. 14pp.

Akbar, P. and Majid, T. (2010).The Effect of ofAflatoxin Levels on Milk production, Reproduction and Lameness in High production Holstein Cows; *Afri.J. of Biotech.*Vol 9 (46), pp. 7905 -7908.

Ananth, S.B and FaridWaliyar (2012).Importance of Aflatoxins in Human and Livestock Health accessed and retrieved at http://www.icrisat.org/aflatoxin/health.asp on 11th August, 2012.

AOAC (2005)Official Methods of Analysis. Association of Official Analytical Chemists, 15th edition., Washington DC, USA.

Awan, J. A. (2001). Fungal Food Poisoning Element of Food Borne Diseases.Institute of Management  and Technology, Enugu State.

Babatunde, B. B andHamzat, R. A. (2005). Effect of Feeding Graded Levels of Kolanut Husk Meal on the Performance of Cockerels. *J. Anim. Prod.* 32 (1): 61 - 66.

Carvajah, M., Rojo, F., Mendez, I and Bolanos, A (2003).Aflatoxin $B_1$ and Its Interconverting Metabolites Aflatoxicol in Milk: The situation in Mexico.*Food Addit. Contamin.* 20, 1077 - 1086.

CAST (Council for Agricultural Science and Technology) (2003). Mycotoxins: Risks in Plant,  Animal and Human Systems. Task Force Report No. 139. Ames Iowa.

Creppy, E. E. (2002).Update of Survey, Regulation and Toxic Effects of Mycotoxin in Europe.*Toxicol.Lett.* 127, 19 - 28.

Dixon, R andCombellas, J. (1983). A Note on Preservation of Wet Brewers' Grains: Inst. Produccion Animal, Facultad de Agronomia, Univ. Central de Venezuela, Maracay, Venezuela; *Tropical-Animal-Production.*, 8: 2, 151-152.

Dohlman, E.(2004). Mycotoxin Regulation; Implications for Agricultural Trade." USDA, 02 October. 2011.

Du Thanh Hang and Preston, T. R. (2005).The Effect of Simple Processing Method of Cassava Leaves on HCN Content and Intake by Growing Pigs. Livestock Research for Rural development. Volume 17, Article No 99.Retrieved from www.cipavorg.co/irrd/irrd1719/hang17099.htm on 7th September, 2005.

Energy Research Centre UNN (2008). Annual Weather Record of University of Nigeria, Nsukka.

Ezedimma, C. I (2001). Relative Factors: Share and Productivity of Horticultural Enterprises in the Urban Enviroment of Lagos. Proc. HORTSON Conference pp. 232.

Fatufe, A.A., Akanbi, I.O., Saba, G.A., Olowofeso, O and Tewe, O. O (2007). Growth Performance and Nutrient Digestibility of Growing Pigs Fed a Mixture of Palm Kernel Meal and Cassava Peel Meal. Livestock Research for Rural Development 19 (12).

Farombi, E. O. (2006). *Afri. J. Biotechnol.* 5(1), 00.1

Gilbert, J and Vargas, E.A (2003). *J. Toxicology* 22: 381.

Ijaiya, A. T., Fasanya, O. O. A and Ayanwale, B. A (2004). Reproductive Performance of Breeding Rabbit Does Fed Maize and Ferment Cassava Peel Meal. *Proceeding of the 29th Annu. Conf. Nig. Soc. Anim.        Prod. (NSAP).*

Jewers, K., Coker, R. D., Blunden, G., Jazwinski, M. J and Sharkey A. J (1986). Problems Involved in the Determination of Aflatoxins levels in Bulk Commodities, *Internat. Biodeterioration*, 22 Pp.83 - 88.

Kuilman, M. E., Maas, R.F.M.,Judah, D. J and Fink-Gremmel,J. (1998).Bovine Hepatic Metabolism of Aflatoxin $B_1$.*J. Agric. Food Chem.* 46: 2707 - 2713.

Lin, L. C.,Liu, F. M., Fu, Y. M and Shih,D. Y. C. (2004). Survey of Aflatoxin $M_1$ Contamination of Dairy Products in Taiwan. *J. Food Drug Anal.*12: 154 -160.

Lu, F. C. (2003). Assessment ofSatety/Risk vs Public Health Concerns:     Aflatoxinsand Hepatocarcinoma. *Enviro.Health Prev. Med.* 7: 235 -238.

Pier, A. C. (1992).Major Biological Consequences of AflatoxicosisinAnimal Production. *J. of Anim. Sci.* 70: 3964 - 3967 accessed and retrieved at http://www.journalofanimalscience.org/content/70/12/3964on 28th August, 2012.

Reddy, C. Vand Farid, W (2012).Properties of Aflatoxin and it Producing Fungi, available at http://www.icrisat.org/aflatoxin/aflatoxin.aspaccessed on 25th July, 2012.

Rhule, S. W. A. (1999).Growth Rate and Carcass Characteristics of Pigsfed on Diets containing Palm Kernel Cake. *Animal Feed Science and Technology* 61: 167-172.

Sashidhare, R. B., Ramakrishna,V and Ramesh, V.B. (1992).Mould and Mycotoxin in Sorghum stored in Traditional Containers in India. *J. Stored Prod. Res.* 28 (4): 257-260.

Serres, H. (1992). Manual of Pig Production in the Tropics. CAB International, Wallingford.

Sun, C, A. and Chen, C. J. (2003). Aflatoxin- induced Hepatocarcinogenesis: Epidemiological Evidence and Mechanistis Considerations. *J. Med. Sci.* 23, 311 - 318.

Odoemelam, S. A and Osu, C. I. (2008). Aflatoxin $B_1$contamination of some edible grains marketed in Nigeria. *E-journal of Chemistry* 6 (2), 308 – 314.

Opadokun, A. (1979). The aflatoxin Content of Locally Consumed Foodstuffs: Part II Sorghum. *Ann. Rep Nig. Stored Prod. Res.* 101.

Ono, E. Y. S., Silva, M., Ribeiro, R. M. R, Ono, M. A., Hayashi, L., Garcia, G.T and Hirooka, E.Y. (2010).Comparison of thin-layer chromatography (TLC), spectrofluorimetry and bright-greenish-yellow fluorescence test for aflatoxin detection in corn.*Braz. Arch .boil. Technol.*Vol 53, No. 3 pp. 687 – 692.

Onyimonyi, A. E. (2002).Nutritional Evaluation of Cassava (*Manihotutilisima* pohl) peel and Bambara (*Voandzoasubtereneat*hower) waste in Pigs diets. Ph.D Thesis, Department of Animal Science, University of Nigeria, Nsukka.

Oyekale, A. S (2009). Climate Variability and its Impact on Agricultural Income and house Welfare in Southern and northern Nigeria. *Electronic Journal of environmental Agricultural and Food Chemistry* 8 (1) 13 - 34.

Ozsoy, S., Altunatmaz, K., Horoz, H., Kasikcle, G., Alkan, S and Bilat, T. (2005).The Relationship between Lameness, Fertility and Aflatoxin in a Dairy Cattle Herd.*Turk. J. Vet. Anim. Sci.* 29: 981-986.

Uddin, I. O (2013). Knowledge, Attitude and Practice of Herbicide Use among farmers in Edo State, Nigeria(M.Sc.) Thesis, Department of Agricultural Extension, University of Nigeria, Nsukka, pp. 46.

Ugwu, S.O.C., Onyimonyi, A. E and Ozonoh, C.I. (2008). Comparative Performance and Haematological Indices of Finishing Broilers fed Palm Kernel Cake, Bambara Offal a nd Rice Husk as Partial Replacement for Maize. *Intl. J. of Poultry Sci.* 7(3): 299-303.

Whitlow, L. (2011).Mycotoxinsand Distiller Grain, North Carolina State University, USA.